Firefighter Essentials:

Elevator Rescue Operations

Zero Mella MD, MHA
Firefighter I and II, NFPA Certified, Virginia Beach
NREMT Certified Emergency Medical Technician
Fire Training Instructor, Olongapo Fire Rescue Team
Disaster Risk Reduction and Management on Health Manager, Olongapo City

Firefighter Essentials: Elevator Rescue Operations
©2023, Zero Mella, Published by Lulu Publishing
All Rights Reserved. No part of this publication should be reproduced, stored in a retrieval system, or transmitted in any form or by any means: electronic, mechanical, photocopying, recording, or otherwise, without the prior written permission of the author and the publisher.

LIMIT OF LIABILITY/DISCLAIMER OF WARRANTY: THE PUBLISHER AND THE AUTHOR MAKE NO REPRESENTATIONS OR WARRANTIES CONCERNING THE ACCURACY OR COMPLETENESS OF THE CONTENTS OF THIS WORK AND SPECIFICALLY DISCLAIM ALL WARRANTIES, INCLUDING WITHOUT LIMITATION WARRANTIES OF FITNESS FOR A PARTICULAR PURPOSE. NO WARRANTY MAY BE CREATED OR EXTENDED BY SALES OR PROMOTIONAL MATERIALS. THE ADVICE AND STRATEGIES CONTAINED HEREIN MAY NOT BE SUITABLE FOR EVERY SITUATION. THIS WORK IS SOLD WITH THE UNDERSTANDING THAT THE PUBLISHER IS NOT ENGAGED IN RENDERING LEGAL, ACCOUNTING, OR OTHER PROFESSIONAL SERVICES. IF PROFESSIONAL ASSISTANCE IS REQUIRED, THE SERVICES OF A COMPETENT PROFESSIONAL PERSON SHOULD BE SOUGHT. NEITHER THE PUBLISHER NOR THE AUTHOR SHALL BE LIABLE FOR DAMAGES ARISING THEREFROM. THE FACT THAT AN ORGANIZATION OR WEBSITE IS REFERRED TO IN THIS WORK AS A CITATION AND/OR A POTENTIAL SOURCE OF FURTHER
INFORMATION DOES NOT MEAN THAT THE AUTHOR OR THE PUBLISHER ENDORSES THE INFORMATION THE ORGANIZATION OR WEBSITE MAY PROVIDE OR RECOMMENDATIONS IT MAY MAKE. FURTHER, READERS SHOULD BE AWARE THAT INTERNET WEBSITES LISTED IN THIS WORK MAY HAVE CHANGED OR DISAPPEARED BETWEEN WHEN THIS WORK WAS WRITTEN AND WHEN IT IS READ.
ISBN 978-1-312-57006-1
Imprint: Lulu.com

Table of Contents

Introduction 8

I. Elevator Rescue: History and Mechanics of Operation 10

    Earliest documented elevator rescue procedures 11

    Early rescue tools and Techniques 12

    Elevator Rescue Procedures Refined and Improved Over Time 12

    Impact of Regulations and Codes 14

II. Mechanics of Elevator Operation 15

    Elevator Operation and Safety Features 15

    Limitations of Elevator Function in Emergency Situations 16

III. Elevator Rescue Procedures 18

    A. Introduction 18

        Legal and ethical considerations 18

        Responsibility of Building Owners and Emergency Responders 18

        Importance of Training and Preparation for Elevator Rescue Operations 19

    B. Overview of Steps and Phases 21

    C. Critical Tasks and Procedures for Each Phase 22

        1. Pre-planning and equipment staging 22

        2. Arrival and Assessment 23

        3. Communication with Occupants 24

        4. Securing the Elevator 25

        5. Gaining Access to the Elevator Car 26

        6. Providing Medical Aid and Extrication 28

        7. Final Steps and Cleanup 29

IV. Response Considerations 31

Factors impacting elevator rescue operations. 31
Situational Awareness and Communication 32
V. Personnel Assignments 34
Overview of key considerations 34
Factors impacting elevator rescue operations 34
    A. Building design and layout: 35
    B. Elevator location and type: 35
    C. Number and condition of occupants: 35
    D. Availability and condition of rescue equipment: 35
Situational Awareness and Communication 35
Overview of Elevator Rescue Personnel Roles and Responsibilities 36
    A. Incident Commander: 36
    B. Elevator Technician: 37
    C. Medical Personnel: 37
    D. Rescue Team Members: 37
Training and Certification Requirements 38
    A. Basic Elevator Rescue Training: 38
    B. Advanced Elevator Rescue Training: 38
    C. Ongoing Training and Refresher Courses: 38
Importance of Teamwork and Coordination 38
Conclusion 39
VI. Phase 1: Preliminary Steps 41
A. Identification of the Stalled Elevator and Initial Assessment 41
B. Communication with the Occupants 42
C. Removal of Occupants from the Stalled Elevator 43

    D. Recycling the Power on the Main Disconnect to the Elevator Car ... 45

    E. Conclusion 45

VII. Phase 2: Lock-Out/Tag-Out 47

    A. Definition of Lock-Out/Tag-Out 47

    B. Procedures for Locking Out/Tagging Out the Elevator Machinery ... 47

        1. Identification of the power sources and how to shut them off: ... 47

        2. Procedures for locking out/tagging out the machinery: 48

        3. Importance of following established safety protocols: 48

    C. Identification of the Power Sources 48

    D. Removal of the Control Key or Lockout Device 49

    E. Testing the Elevator Controls 50

    E. Conclusion 51

VIII. Phase 3: Car Movement and Exit 52

    A. Procedures for Movement of the Stalled Elevator Car 52

        1. Explanation of the procedures for moving the stalled elevator car: 52

        2. Importance of ensuring safety during the movement process: 52

        3. Techniques for controlling the movement of the elevator car: 52

    B. Preparation of the Elevator Hoistway 53

    Procedures for preparing the elevator hoistway for the movement of the car. 53

    Importance of verifying that the hoistway is safe for movement 53

    C. Methods for Accessing the Elevator Hoistway 54

    D. Exit Procedures for the Occupants 54

    Procedures for safely and efficiently removing occupants from the elevator car 54

- Importance of prioritizing occupant safety during the exit process . 55
- Techniques for managing occupants with medical conditions or injuries 55
- E. Final Assessment and Recovery of Equipment.................................... 55
  - Procedures for assessing the elevator machinery and Hoistway 55
  - Recovery of equipment 56
- F. Conclusion.. 56

IX. Firefighter Training Guidelines 57
- A. Recruit Firefighter Training Guidelines... 58
- B. In-House Firefighter Training Guidelines.. 59
- C. Conclusion 60

X. Debriefing Procedures for Firefighters... 62
- A. Importance of Debriefing Procedures... 63
- B. Guidelines for Effective Debriefing Sessions 64
  1. Create a safe and supportive environment for debriefing: 64
  2. Facilitate open and honest communication among firefighters: .. 64
  3. Identify areas for improvement and implement changes: 64
- C. Conclusion 65

XI. Codes Related to Elevator Rescue.. 66
- A. NFPA Codes Related to Elevator Rescue 67
  1. Explanation of the National Fire Protection Association (NFPA) ... 67
  2. Overview of the NFPA codes related to elevator rescue 67
  3. Explanation of the specific requirements outlined in the NFPA codes related to elevator rescue... 69
- B. International Guidelines Related to Elevator Rescue......................... 70
- C. Conclusion 73

About the Author 75

Introduction

Welcome to " *Firefighter Essentials: Elevator Rescue Operations*," a comprehensive guide that delves into the critical realm of saving lives during elevator emergencies. Authored by Zero Mella, a dedicated physician and fire training instructor from the Philippines, this book combines his vast experience in healthcare and firefighting to provide invaluable insights and knowledge to firefighters and first responders worldwide.

Elevator rescue operations pose unique challenges, requiring specialized skills, precise techniques, and a thorough understanding of safety protocols. The gravity of these situations emphasizes the need for a comprehensive resource that equips firefighters with the necessary tools to navigate through complex scenarios with confidence and competence.

In this book, Zero Mella draws upon his firsthand experiences as a practicing physician and a fire training instructor, highlighting the importance of proper training, efficient communication, and adherence to established codes and guidelines. Each chapter is meticulously crafted to address the specific phases of elevator rescue operations, from initial response and assessment to car movement, occupant exit procedures, and debriefing sessions.

"Ultimate Firefighter: Elevator Rescue Operations" begins by establishing a foundation of knowledge regarding elevator systems, their components, and potential hazards. It then progresses into an exploration of the key phases of elevator rescue operations, providing step-by-step procedures, safety considerations, and best practices to ensure the safety of both occupants and responders.

This book is designed as a practical and accessible resource, presenting clear explanations, diagrams, and real-life case studies to enhance understanding and facilitate effective learning. Zero Mella's expertise and dedication to serving the least fortunate Filipinos further emphasize the importance of aiding those in need, even in the most challenging circumstances.

Whether you are a firefighter, a first responder, or someone interested in elevating your knowledge of elevator rescue operations, "Ultimate Firefighter: Elevator Rescue Operations" will serve as an indispensable companion in your quest for proficiency and excellence. By following the guidelines presented within these pages, you will be equipped with the skills and confidence to save lives and mitigate risks effectively.

Embrace the opportunity to enhance your expertise, advance your career, and make a lasting impact in the field of firefighting. Let "Ultimate Firefighter: Elevator Rescue Operations" be your trusted guide, illuminating the path towards safer and more successful elevator rescue operations.

I. Elevator Rescue: History and Mechanics of Operation

The history of elevator technology is closely tied to the history of the Industrial Revolution, which brought about significant changes in manufacturing and transportation technologies. In the early 1800s, crude, steam-powered elevators were first developed, and by the mid-1800s, hydraulic and electric elevators had begun to emerge. These early elevators were relatively slow and had limited capacity, but they represented a significant technological advancement at the time.

Over the years, technological advancements have continued to improve the design and functionality of elevators. Today, elevators are faster, more efficient, and safer than ever before, with features like automatic doors, computerized controls, and sophisticated safety systems. However, these same advancements have also introduced new challenges for firefighters responding to elevator emergencies. For example, modern elevators may use complex electronic systems that can be difficult to disable in an emergency, and some elevators can travel at speeds of up to 40 miles per hour, which can make rescue operations more difficult and dangerous. As a result, firefighters must be highly trained and equipped with specialized tools and techniques to respond to elevator emergencies safely and effectively.

Modern elevators are designed with state-of-the-art technology that ensures safe and efficient transportation of passengers. The advanced control systems in modern elevators are responsible for monitoring all elevator functions and controlling the movement of the elevator car. These control systems work in conjunction with an array of sensors that are strategically placed throughout the elevator to detect weight, speed, and movement.

One of the most critical safety features in modern elevators is the emergency braking system. This system is designed to detect any malfunction in the elevator and automatically stop the car. The emergency brakes are typically located on the elevator's main drive system, and they engage when an emergency stop button is pushed or when sensors detect that the elevator is moving too quickly.

Despite the many safety features in modern elevators, they can still pose challenges for firefighters during rescue operations. For instance, firefighters must be able to accurately assess the situation, understand the elevator's design and capabilities, and work quickly and efficiently to rescue stranded occupants. Additionally, firefighters must be prepared to navigate through cramped spaces, deal with potential hazards like electrical wires or hydraulic fluids, and avoid any sudden movements that could trigger the elevator's safety mechanisms.

Overall, while modern elevators have undoubtedly improved in terms of safety, they have also introduced new complexities that require specialized knowledge and training to address effectively during rescue operations.

Understanding the evolution of elevator technology is crucial for firefighters responding to elevator emergencies. It enables them to develop effective strategies and procedures for dealing with the unique challenges posed by modern elevators. By understanding the technology behind elevators, firefighters can more effectively assess the situation, select the appropriate tools and techniques, and execute the necessary tasks to safely rescue occupants.

Earliest documented elevator rescue procedures

As the number of high-rise buildings increased, so did the number of elevator emergencies. As a result, firefighters developed new rescue techniques and tools to deal with these emergencies. In the early 1900s, firefighters began using special elevator cages to safely transport occupants from stalled elevators. These cages were designed to be attached to the top of the elevator car, and occupants could then be lifted out of the car and transported to safety.

As elevators became more sophisticated, firefighters developed new rescue techniques to keep pace with the changes. For example, in the 1960s and 1970s, firefighters began using hydraulic jacks to pry open elevator doors, allowing them to access stalled cars. They also developed specialized tools, such as elevator keys and door wedges, to help them gain access to the elevator and safely rescue occupants.

Today, firefighters are equipped with a wide range of specialized tools and equipment designed specifically for elevator rescue operations. These tools include elevator-specific keys, door wedges, and hydraulic prying tools, as well as portable elevators and hoists that can be used to transport occupants from stalled elevators. Firefighters also receive specialized training in elevator rescue techniques and are equipped with the knowledge and skills necessary to safely rescue occupants from elevators in emergencies.

Early rescue tools and techniques

As elevator technology advanced, so did the tools and techniques used by firefighters to rescue occupants from stalled elevators. In the late 1800s, elevator cages were developed to provide a safer way to rescue occupants from stalled elevator cars. These cages were equipped with ropes or chains that could be lowered into the elevator shaft, allowing occupants to climb out of the stalled car and into the cage. Firefighters would then hoist the cage to the nearest floor and evacuate the occupants.

In the early 1900s, manually operated elevators were developed, providing firefighters with a safer and more efficient way to rescue occupants from stalled elevators. These elevators were equipped with a hand crank that could be used to raise and lower the elevator car, allowing firefighters to rescue occupants without having to climb up and down the elevator shaft.

Overall, the historical development of elevator rescue procedures has been driven by the need to provide a safe and effective means of rescuing occupants from stalled elevators. As elevator technology continues to advance, so will the tools and techniques used by firefighters to respond to elevator emergencies.

Elevator Rescue Procedures Refined and Improved Over Time

The evolution of elevator technology has brought about significant changes in the safety features of elevators. Elevators now have various safety features such as door interlocks, emergency lighting, communication systems, and automatic control systems, among others. These safety features have made elevators safer for passengers but have also increased the complexity of rescue procedures for firefighters.

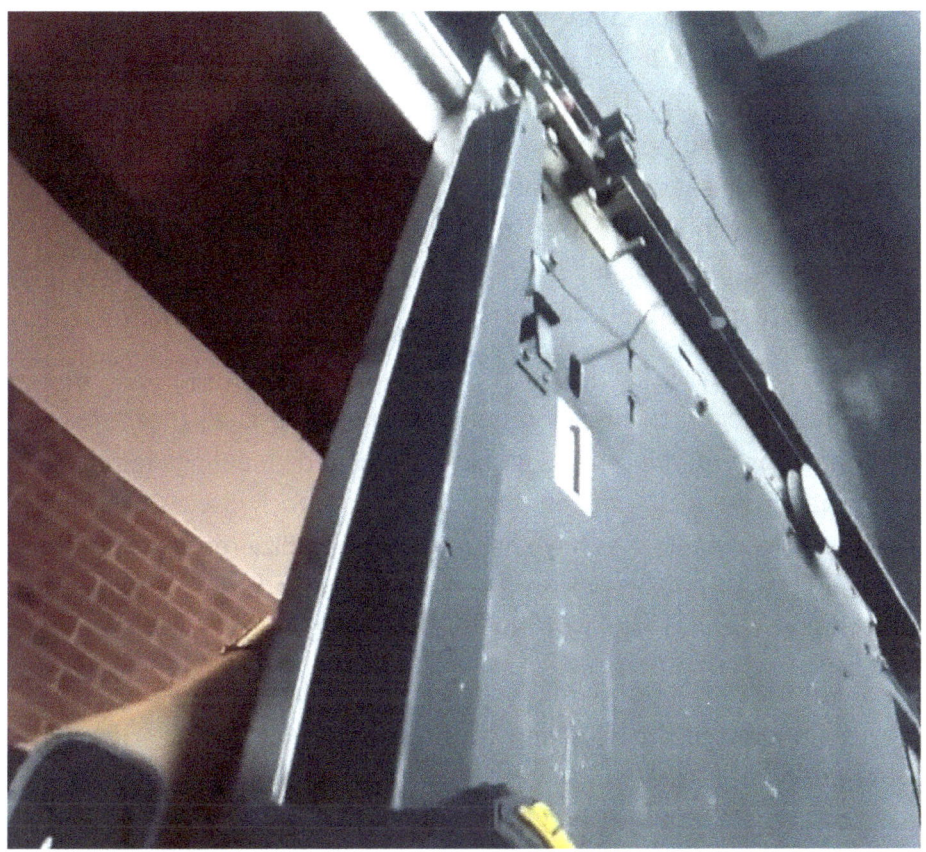

Figure 1: Firefighter Training using an elevator key. The view is from the inside of an elevator door with the interlock mechanism on the top of the image.

With the advancements in elevator technology, elevator-specific rescue tools and equipment have been developed to assist firefighters in rescue operations. Some of these tools and equipment include the elevator door key, elevator access tools, elevator emergency operations panels, and elevator communication devices. These tools and equipment are essential for firefighters to gain access to the elevator, communicate with the occupants, and safely evacuate them in case of an emergency.

Impact of Regulations and Codes

Regulatory bodies, such as the National Fire Protection Association (NFPA) and the Occupational Safety and Health Administration (OSHA), have played a critical role in establishing safety standards for elevators. For example, the NFPA 101 Life Safety Code provides guidelines for elevator safety and evacuation procedures in the event of an emergency. Similarly, OSHA has established regulations under 29 CFR 1910.36 for elevator operation and maintenance to ensure the safety of workers and occupants. These standards are designed to ensure the safe operation of elevators and reduce the risk of accidents and emergencies.

The safety standards established by regulatory bodies have had a significant impact on the development of elevator rescue procedures and equipment. For example, the NFPA has established guidelines for elevator rescue operations, including the three-phase approach discussed earlier in this guide. These guidelines have helped to standardize rescue procedures and ensure that firefighters have the knowledge and tools they need to respond to elevator emergencies safely and effectively.

In addition to setting standards for rescue procedures, regulatory bodies have also influenced the development of rescue equipment. For example, OSHA requires that elevator shafts be equipped with ladderways or other means of access for emergency responders. This requirement has led to the development of specialized rescue ladders and harnesses that allow firefighters to safely access elevator shafts.

Overall, the influence of regulatory bodies has been instrumental in shaping the way firefighters respond to elevator emergencies. By establishing safety standards and guidelines, these organizations have helped to ensure that firefighters are equipped with the knowledge and tools they need to respond to emergencies safely and effectively.

II. Mechanics of Elevator Operation

Hydraulic elevators, traction elevators, and machine-room-less elevators are the three main types of elevators. Hydraulic elevators are powered by a hydraulic piston that lifts the elevator car. These types of elevators are typically used in buildings with fewer than six floors. Traction elevators, on the other hand, are powered by an electric motor that drives a cable and pulley system. These elevators are commonly found in mid-rise and high-rise buildings. Machine-room-less elevators are like traction elevators but do not require a separate machine room to operate.

In addition to the different types of elevators, various mechanical systems power them. Cables and pulleys are commonly used in traction elevators, while hydraulic elevators rely on a hydraulic system to lift the elevator car. In some cases, elevators may use a combination of these systems. For example, a traction elevator may use a hydraulic system to power the emergency brakes. Understanding the mechanical systems that power elevators are essential for firefighters who may need to disable or control these systems during a rescue operation.

Elevator Operation and Safety Features

Elevators operate using a system of cables, pulleys, and motors that allow the elevator cab to move up and down between floors. The cab is suspended from steel cables that are attached to a counterweight on one end and a drive sheave on the other. The drive sheave is powered by an electric motor, which controls the speed and direction of the elevator cab.

In the event of an emergency, elevators are equipped with several safety features to protect passengers and first responders. One of the most important safety features is the emergency braking system, which is designed to automatically stop the elevator cab in the event of a malfunction or power outage. This system typically consists of a mechanical brake that engages when the elevator loses power or when an emergency stop button is pressed.

Elevators also feature communication and alarm systems that allow passengers to call for help and provide information to first responders. In many cases, elevators are equipped with intercoms or telephones that allow passengers to communicate directly with building security or maintenance staff. Some elevators are also equipped with alarm buttons that can be pressed in the event of an emergency, which will activate a loud alarm and alert building occupants and first responders to the situation.

Limitations of Elevator Function in Emergency Situations

Figure 2: Elevator technician resetting an elevator using elevator keys.

Elevators are designed to operate under normal conditions, but they can experience malfunctions or failures that require emergency response. Understanding the limitations of elevator function in emergencies is essential for developing effective rescue strategies.

One common limitation of elevators is power failure. In the event of a power outage, elevators can become stuck between floors, trapping occupants inside. In some cases, elevators may automatically stop at the nearest floor and open the doors to allow occupants to exit, but this is not always the case. Additionally, elevators may not operate during power outages at all, leaving occupants stranded until power is restored or they can be rescued.

Another limitation is a mechanical malfunction. Elevator components can wear down or fail over time, leading to malfunctions such as doors that fail to open or close properly, cables that snap or become loose, or brakes that fail. In some cases, elevators may also experience software or control system malfunctions that cause them to behave erratically or stop functioning altogether.

Elevator design and age can also impact rescue operations. Older elevators may not have the same safety features as newer ones, making them more dangerous for occupants and more challenging for firefighters to rescue. Additionally, elevators may be designed differently depending on their intended use, such as those in high-rise buildings versus those in low-rise buildings, which can affect rescue strategies.

Understanding the limitations of elevator function in emergencies is critical for developing effective rescue strategies. Firefighters must be trained to assess the situation quickly and determine the best course of action, whether that involves attempting to restart the elevator, using specialized rescue equipment, or calling in a specialized elevator rescue team.

III. Elevator Rescue Procedures

A. Introduction

Legal and ethical considerations

Elevator rescue operations are subject to a range of laws and regulations that are designed to ensure the safety of elevator occupants and emergency responders. These regulations include NFPA 130: Standard for Fixed Guideway Transit and Passenger Rail Systems, OSHA standards, and local building codes.

Building owners and managers have a responsibility to ensure the safety of their elevators and occupants. This includes regular maintenance, inspections, and testing of safety features. Failure to do so can result in legal liability if an accident or emergency occurs.

Ethical considerations also come into play during elevator rescue operations. Emergency responders have to rescue occupants who are stranded in elevators, but they must also balance this duty with the need to protect their safety and the safety of other occupants.

Rescuers must also consider the potential risks and consequences of their actions, as well as the impact on the occupants they are rescuing. For example, rushing to rescue occupants from a stalled elevator without proper planning or equipment can result in injuries or other complications. Therefore, proper training and preparation for elevator rescue operations are essential to ensure that emergency responders can carry out their duties effectively and safely.

Responsibility of Building Owners and Emergency Responders

As stated above, building owners have the responsibility of ensuring that elevators are regularly maintained and inspected to ensure safe and efficient operation. They must also comply with local building codes and safety regulations related to elevators. In the event of an elevator emergency, building owners must promptly notify emergency responders and provide access to the elevator machinery room and control systems.

On the other hand, emergency responders have the responsibility of responding quickly and efficiently to elevator emergencies to ensure the safety of occupants. They must be properly trained and equipped with the necessary tools and equipment to carry out elevator rescue operations. Additionally, they must work closely with building owners to gather critical information about the elevator and the emergency, such as the location of the elevator car, the number of occupants, and any potential hazards.

Cooperation and communication between building owners and emergency responders are crucial in responding to elevator emergencies. Building owners can provide important information about the elevator's mechanical systems, while emergency responders can offer insights into the specific challenges of elevator rescue operations. By working together, they can develop effective strategies to safely rescue occupants and minimize the risk of injury or damage.

Importance of Training and Preparation for Elevator Rescue Operations

Elevator rescue operations require specialized skills and knowledge to ensure the safety of both occupants and responders. Therefore, regular training and preparation are critical to effectively respond to elevator emergencies. In this section, we will discuss the necessary skills and knowledge required for elevator rescue operations, the importance of regular training and drills, and the benefits of establishing standard operating procedures.

1. Skills and Knowledge Required for Elevator Rescue Operations

Elevator rescue operations require specialized skills and knowledge, such as:

- Knowledge of elevator types, mechanical systems, and safety features

- Familiarity with elevator rescue equipment, such as emergency stop switches, hoists, and safety harnesses
- Understanding of rescue procedures and protocols
- Ability to communicate effectively with occupants, other responders, and building owners.
- Basic medical training to provide aid to injured occupants.

In addition, responders must have the physical ability to climb ladders, operate heavy equipment, and work in confined spaces.

2. Importance of Regular Training and Drills

Regular training and drills are essential to maintain readiness and familiarity with elevator rescue procedures and equipment. Responders should receive regular training on elevator types, mechanical systems, safety features, and rescue protocols. They should also receive hands-on training in the use of rescue equipment and practice simulated rescue scenarios.

Drills should be conducted regularly to ensure that responders are prepared to handle emergencies effectively. The drills should include all phases of elevator rescue operations, from initial response to post-rescue procedures.

3. Benefits of Establishing Standard Operating Procedures

Standard operating procedures (SOPs) for elevator rescue operations provide a consistent and effective framework for responding to elevator emergencies. SOPs outline the steps and procedures that responders should follow in different types of elevator emergencies, ensuring that they respond appropriately and safely.

SOPs should cover all phases of elevator rescue operations, from initial response to post-rescue procedures. They should also include guidelines for communication with occupants, coordination with other responders, and the use of rescue equipment.

In addition to providing a consistent framework for responding to elevator emergencies, SOPs also help to identify areas where improvements can be made in rescue procedures and equipment. Regular review and updating of SOPs are necessary to ensure that they remain relevant and effective.

B. Overview of Steps and Phases

Elevator rescue operations involve several steps and phases, each of which is critical to the success and safety of the operation. The following is an overview of the typical steps and phases involved in elevator rescue operations:

1. **Pre-planning and equipment staging** Pre-planning involve developing strategies and procedures for responding to elevator emergencies. This includes identifying potential hazards and obstacles, mapping out the building's elevator system, and establishing lines of communication with building management and occupants. Equipment staging involves ensuring that all necessary equipment, such as elevator keys, rescue kits, and communication devices, are readily available and in good working condition.

2. **Arrival and assessment**: Upon arrival at the scene, the priority is to assess the situation and determine the number and condition of occupants in the stalled elevator. This involves communication with occupants, as well as an assessment of the elevator's location and mechanical condition.

3. **Communication with occupants**: Effective communication with occupants is critical in ensuring their safety and well-being during the rescue operation. This involves explaining the rescue procedures, providing reassurance, and obtaining necessary medical and contact information.

4. **Securing the elevator**: Once the occupants have been identified and assessed, the elevator must be secured to prevent unintended movement. This involves activating the emergency brake system and other safety features.

5. **Gaining access to the elevator car**: Depending on the location and condition of the elevator, gaining access to the car may involve using elevator keys, accessing the elevator shaft, or cutting through the elevator doors. This step requires careful planning and execution to ensure the safety of both occupants and rescuers.

6. **Providing medical aid and extrication**: Once access has been gained, rescuers can provide medical aid to occupants as needed and begin the process of extricating them from the elevator car. This involves carefully maneuvering occupants through the elevator doors and shaft and may require specialized equipment and techniques.

7. **Final steps and cleanup**: Once all occupants have been safely removed from the elevator, the rescue operation is complete. Rescuers must then ensure that all equipment is accounted for and that the elevator is returned to normal operation. Any injuries or damage must be documented and reported as necessary.

Effective execution of each of these steps and phases requires careful planning, communication, and coordination between building owners, occupants, and emergency responders.

C. Critical Tasks and Procedures for Each Phase

1. Pre-planning and equipment staging

In the pre-planning and equipment staging phase, it is essential to take the necessary steps to ensure that the rescue operation proceeds smoothly and safely. Some critical tasks and procedures for this phase include:

> **a. Identifying the type of elevator and its location**: The first step in any elevator rescue operation is to determine the type of elevator and its location within the building. This information is critical as it will help emergency responders to identify the appropriate rescue equipment and techniques required for the specific type of

elevator. For example, hydraulic elevators require different tools and procedures compared to traction elevators.

b. **Assessing the building's emergency power supply and communication systems:** In the event of a power failure, it is crucial to ensure that the building's emergency power supply is functioning correctly, and communication systems are operational. This assessment includes checking the availability of backup generators and the functionality of emergency communication devices such as phones and two-way radios.

c. **Identifying potential hazards and developing a rescue plan**: The pre-planning phase should also include identifying any potential hazards that may impede the rescue operation, such as the presence of hazardous materials or the risk of fire. Once these hazards have been identified, emergency responders should develop a rescue plan that considers these potential hazards and outlines the necessary steps and procedures required to mitigate any risks. This plan should also include the necessary equipment and personnel required to safely carry out the rescue operation.

2. Arrival and Assessment

When firefighters arrive at the scene of an elevator emergency, the first step is to locate the elevator and assess the situation. This phase of the rescue operation is critical because it provides the necessary information to formulate an effective rescue plan. The following are the critical tasks and procedures for the arrival and assessment phase:

a. **Determining the elevator's location and condition**: The first task is to determine the location of the stalled elevator and its condition. This information can be obtained from the building's security personnel or elevator maintenance personnel. If the elevator is equipped with a communication system, the firefighters can use it to communicate with the occupants and gather additional information.

b. Verifying the number and condition of occupants: The next task is to verify the number and condition of occupants in the stalled elevator. This information will help firefighters to determine the severity of the situation and the resources required for the rescue operation. If the elevator has a monitoring system, the firefighters can use it to identify the number of occupants and their condition. If not, they can communicate with the occupants to obtain the information.

c. Assessing the severity of any injuries or medical conditions: Once the number and condition of occupants are verified, the firefighters will assess the severity of any injuries or medical conditions. If any occupants require medical attention, the firefighters will provide initial medical aid until the paramedics arrive. The firefighters will also determine if any occupants require immediate extrication or if they can safely wait for the elevator to be repaired or reset.

3. Communication with Occupants

During an elevator emergency, communication with stranded occupants is a critical task. The following are the procedures and considerations involved in communicating with occupants:

a. Establishing contact with occupants and providing reassurance: The first step is to establish contact with the occupants and reassure them that help is on the way. This can be done through the emergency communication system in the elevator, such as the intercom or emergency phone. It is important to speak calmly and clearly and to provide accurate information on the rescue operation's progress.

b. Obtaining information on medical conditions and injuries: Once contact is established, the rescuers should obtain information on any medical conditions or injuries that the occupants may have.

This information will help determine the priority of the rescue operation and the type of medical assistance required.

c. **Providing instructions on how to prepare for rescue**: The rescuers should provide instructions to the occupants on how to prepare for rescue, such as sitting or lying down to reduce the risk of injury during the rescue. The occupants should also be instructed to conserve their energy and avoid unnecessary movement to prevent the elevator from swaying or shifting.

The communication phase is crucial to the success of the rescue operation, and it requires clear communication and effective coordination between the rescuers and the occupants. Proper communication can help reduce panic and anxiety and make the rescue operation safer and more efficient.

4. Securing the Elevator

When securing the elevator, it is critical to ensure that the elevator car is stable and not moving before proceeding with rescue operations. The following are critical tasks and procedures for securing the elevator:

a. **Isolating power to the elevator:** The first step in securing the elevator is to isolate power to the elevator. This can be accomplished by shutting off the main power supply to the elevator or by using an elevator control switch located in the machine room.

b. **Securing the doors and preventing movement of the car**: Once the power to the elevator has been shut off, the doors to the elevator should be secured to prevent movement of the car. This can be done by using elevator door wedges, which prevent the doors from closing, or by using elevator door clamps, which prevent the doors from opening.

c. **Setting up ventilation and lighting as necessary:** Depending on the situation, it may be necessary to set up ventilation and lighting inside the elevator car to ensure that occupants have fresh air and can see what is happening during the rescue operation. Portable ventilation and lighting equipment should be available for use in emergencies.

5. Gaining Access to the Elevator Car
a. Determining the safest and most efficient access point.

Once the assessment of the elevator's location and condition is made, it's important to determine the safest and most efficient access point for the rescue operation. The following steps can be taken:

1. **Evaluate the location of the elevator**: The location of the elevator can affect the access point choice. For instance, if the elevator is in a high-rise building, access from above may be necessary.

2. **Assess the elevator's condition**: Consider the condition of the elevator and whether there is any structural damage that may affect the choice of access point.

3. **Determine whether it is safe to access the elevator car from above or below**: Depending on the situation, it may be safer to access the elevator car from above or below. For example, if the elevator is stuck between floors, accessing it from above may be the best option.

4. **Consider the type of elevator and its design**: Different types of elevators have different designs and may require different access points. For example, hydraulic elevators may have an access panel on the side, while traction elevators may have a hatch on the roof.

5. **Evaluate the size and weight of equipment needed for access**: Depending on the chosen access point, equipment such as ladders, ropes, or harnesses may be required to gain access to the elevator car. The size and weight of this equipment should be taken into consideration when determining the access point.

Once the access point has been determined, the rescue team can proceed with gaining access to the elevator car. It is important to follow proper safety procedures to avoid injury to rescuers or occupants.

b. Using elevator-specific rescue tools and equipment.

In addition to the tools and equipment mentioned above, there are also specialized elevator rescue devices that can be used to aid in the extraction of occupants. These devices include evacuation chairs, harnesses, and slings.

When using elevator-specific rescue tools and equipment, it is important to use them in a way that minimizes the risk of injury to the rescuers or occupants. This may involve using protective gear, such as gloves or eye protection, and following proper lifting techniques to avoid strains or other injuries.

It is also important to be aware of any potential hazards or obstacles in the elevator car that may impede the rescue operation. For example, there may be debris or sharp objects that could cause injury, or there may be electrical or mechanical components that need to be avoided. Rescuers should take the time to carefully assess the situation and plan their approach accordingly.

c. Following proper safety procedures to avoid injury to rescuers or occupants.

1. Prioritize the safety of all individuals involved in the rescue operation, including rescuers, occupants, and bystanders.

2. Use caution when entering the elevator car and be prepared for any unexpected movements or conditions. If the elevator is not level with the floor, consider using a ladder or other equipment to safely enter the car.

3. Be aware of any potential hazards, such as sharp edges or exposed electrical components, and take appropriate measures to mitigate risks.

4. If possible, secure the elevator car to prevent any unexpected movements that could cause injury or further harm to occupants.

5. Maintain constant communication with other rescuers and occupants to ensure everyone is aware of the rescue plan and any potential risks or hazards.

6. Follow proper lifting and carrying techniques to avoid injury to rescuers or occupants during extrication.

7. Ensure that occupants are secured and safely removed from the elevator car before leaving the rescue site and provide any necessary medical attention or care.

By following these critical tasks and procedures, rescuers can gain access to the elevator car safely and efficiently, minimizing the risk of injury to themselves and the occupants.

6. Providing Medical Aid and Extrication

a. Assessing and treating injuries or medical conditions.

Once access to the elevator car is gained, assess the occupants' injuries or medical conditions. Provide appropriate medical aid such as CPR, first aid, or the use of an automated external defibrillator (AED) as needed. If necessary, call for additional medical assistance or transport to a medical facility.

b. Using appropriate equipment and techniques to remove occupants from the elevator car.

Determine the best method to remove occupants from the elevator car based on their medical condition, the type of elevator, and the available equipment. Use appropriate rescue tools and equipment to remove occupants safely and efficiently from the elevator car. Ensure that all occupants are safely extricated and transported to a safe area.

c. Providing continued care and transport to medical facilities.
Provide continued medical care and monitoring to occupants as necessary. Coordinate with emergency medical services (EMS) for transport to a medical facility. Provide documentation of the rescue and medical care provided to the occupants for medical records and liability purposes.

7. Final Steps and Cleanup

Once all occupants are safely removed from the elevator, it is important to take steps to secure the elevator and complete the necessary paperwork to document the rescue operation. The following are critical tasks and procedures for the final steps and cleanup phase:

a. Ensuring all occupants are safely out of the elevator.
- Verify that all occupants have been safely removed from the elevator car and are receiving any necessary medical attention.
- Confirm that no remaining individuals are in the elevator shaft or machinery room.

b. Removing equipment and securing the elevator.
- Collect all rescue equipment and tools, ensuring that everything is properly stowed and secured.
- Follow proper procedures for securing the elevator and isolating power, following manufacturer instructions and local regulations.
- Communicate with building management or elevator maintenance personnel to arrange for any necessary repairs or inspections.

c. Completing incident reports and debriefing with all involved parties.
- Prepare and submit incident reports as local regulations or departmental policy require.
- Conduct a debriefing session with all personnel involved in the rescue operation to discuss any lessons learned or areas for improvement.

- Review the rescue plan and make any necessary updates or revisions based on the experience gained from the operation.

By completing these critical tasks and procedures, emergency responders can ensure that the rescue operation is safely and effectively concluded. It is important to prioritize the safety and well-being of all individuals involved, including both the occupants and emergency responders, throughout each phase of the rescue operation.

IV. Response Considerations

In any elevator emergency, the safety of the occupants and responders should always be the top priority. Time is of the essence, and a quick and efficient response is necessary to prevent further injury or harm.

Following established rescue procedures and protocols is crucial in ensuring that the rescue operation is conducted safely and effectively. This includes assessing the elevator's location and condition, identifying potential hazards, and determining the best access point and rescue tools to use.

Additionally, communication and situational awareness are key considerations during elevator rescue operations. Effective communication between responders and occupants can help alleviate anxiety and ensure everyone is on the same page regarding the rescue plan. Situational awareness is also essential to identify any new hazards or changing conditions that may arise during the rescue operation.

Other key considerations include the building's design and elevator location, as well as the number and condition of occupants involved in the emergency. These factors can significantly impact the rescue operation and must be considered when developing a rescue plan.

Overall, prioritizing safety, following established protocols, and maintaining effective communication and situational awareness are critical considerations in any elevator emergency.

Factors impacting elevator rescue operations.

Elevator rescue operations can be impacted by a variety of factors that rescuers should consider when responding to an emergency. Some of the key factors that can impact elevator rescue operations include:

1. **Building design and layout:** The design and layout of the building can affect the accessibility of the elevator and the location of the

access points for rescue. For example, buildings with multiple elevators or elevator banks may present challenges in identifying the location of the affected elevator. Buildings with restricted access or difficult-to-reach locations can also add complexity to the rescue operation.

2. **Elevator location and type:** The location and type of elevator can impact the access points, tools, and techniques required for rescue. Elevators located in confined spaces or at high elevations may require specialized equipment and training. Different types of elevators, such as hydraulic or traction elevators, may require different rescue techniques.

3. **Number and condition of occupants:** The number and condition of occupants in the elevator can affect the urgency and complexity of the rescue operation. The presence of injured or incapacitated occupants may require immediate medical attention and specialized extrication techniques.

4. **Availability and condition of rescue equipment**: The availability and condition of rescue equipment can impact the efficiency and safety of the rescue operation. Rescuers should ensure that all equipment is properly maintained and readily available and that they are familiar with its use.

By considering these factors, rescuers can develop an effective plan and response to an elevator emergency.

Situational awareness and communication

Situational awareness and effective communication are crucial during elevator rescue operations. Responders must maintain constant assessment and monitoring of the rescue environment to identify potential hazards and risks. By recognizing these hazards, they can take the necessary precautions to prevent further harm to occupants and responders.

Clear and concise communication between all parties involved in the rescue operation is essential to prevent confusion or accidents. Responders must establish clear roles and responsibilities for each member of the rescue team, including the incident commander, elevator mechanic, and medical personnel.

The incident commander is responsible for the overall management and coordination of the rescue operation, including communication with building management and emergency services. The elevator mechanic is responsible for ensuring the elevator is safely secured and providing technical expertise on elevator systems. Medical personnel provides care and treatment to any injured occupants.

Effective communication also requires the use of standard terminology and procedures to ensure everyone understands the information being shared. The incident commander should establish a communication plan and ensure all responders are familiar with it.

In summary, situational awareness and effective communication are critical to the success of elevator rescue operations. Responders must remain vigilant to potential hazards and risks while maintaining clear and concise communication to ensure a safe and efficient rescue.

V. Personnel Assignments
Overview of key considerations

In any elevator emergency, the safety of the occupants and responders should always be the top priority. Time is of the essence, and a quick and efficient response is necessary to prevent further injury or harm.

Following established rescue procedures and protocols is crucial in ensuring that the rescue operation is conducted safely and effectively. This includes assessing the elevator's location and condition, identifying potential hazards, and determining the best access point and rescue tools to use.

Additionally, communication and situational awareness are key considerations during elevator rescue operations. Effective communication between responders and occupants can help alleviate anxiety and ensure everyone is on the same page regarding the rescue plan. Situational awareness is also essential to identify any new hazards or changing conditions that may arise during the rescue operation.

Other key considerations include the building's design and elevator location, as well as the number and condition of occupants involved in the emergency. These factors can significantly impact the rescue operation and must be considered when developing a rescue plan.

Overall, prioritizing safety, following established protocols, and maintaining effective communication and situational awareness are critical considerations in any elevator emergency.

Factors impacting elevator rescue operations.

Elevator rescue operations can be impacted by a variety of factors that rescuers should consider when responding to an emergency. Some of the key factors that can impact elevator rescue operations include:

A. Building design and layout:

The design and layout of the building can affect the accessibility of the elevator and the location of the access points for rescue. For example, buildings with multiple elevators or elevator banks may present challenges in identifying the location of the affected elevator. Buildings with restricted access or difficult-to-reach locations can also add complexity to the rescue operation.

B. Elevator location and type:

The location and type of elevator can impact the access points, tools, and techniques required for rescue. Elevators located in confined spaces or at high elevations may require specialized equipment and training. Different types of elevators, such as hydraulic or traction elevators, may require different rescue techniques.

C. Number and condition of occupants:

The number and condition of occupants in the elevator can affect the urgency and complexity of the rescue operation. The presence of injured or incapacitated occupants may require immediate medical attention and specialized extrication techniques.

D. Availability and condition of rescue equipment:

The availability and condition of rescue equipment can impact the efficiency and safety of the rescue operation. Rescuers should ensure that all equipment is properly maintained and readily available and that they are familiar with its use.

By considering these factors, rescuers can develop an effective plan and response to an elevator emergency.

Situational awareness and communication

Situational awareness and effective communication are crucial during elevator rescue operations. Responders must maintain constant assessment and monitoring of the rescue environment to identify potential hazards and risks. By recognizing these hazards, they can take the necessary precautions to prevent further harm to occupants and responders.

Clear and concise communication between all parties involved in the rescue operation is essential to prevent confusion or accidents. Responders must establish clear roles and responsibilities for each member of the rescue team, including the incident commander, elevator mechanic, and medical personnel.

The incident commander is responsible for the overall management and coordination of the rescue operation, including communication with building management and emergency services. The elevator mechanic is responsible for ensuring the elevator is safely secured and providing technical expertise on elevator systems. Medical personnel provides care and treatment to any injured occupants.

Effective communication also requires the use of standard terminology and procedures to ensure everyone understands the information being shared. The incident commander should establish a communication plan and ensure all responders are familiar with it.

In summary, situational awareness and effective communication are critical to the success of elevator rescue operations. Responders must remain alert to potential hazards and risks while maintaining clear and concise communication to ensure a safe and efficient rescue.

Ooverview of Elevator Rescue Personnel Roles and Responsibilities

Elevator rescue operations require coordinated efforts from multiple personnel with specialized skills and expertise. The following are the main roles and responsibilities of elevator rescue personnel:

A. Incident Commander:

The incident commander is responsible for overseeing and managing the rescue operation. This person is responsible for establishing an incident command system, ensuring that all personnel are following established protocols, and making decisions that prioritize the safety of all involved parties. The incident commander also serves as the primary liaison between rescue personnel and building management or emergency services.

B. Elevator Technician:

The elevator technician is responsible for assessing the condition of the elevator, determining the safest access point for the rescue, and using specialized tools and equipment to gain access to the elevator car. The technician may also assist with securing the elevator and ensuring that it is properly shut down.

C. Medical Personnel:

Medical personnel, such as EMTs or paramedics, are responsible for assessing the medical condition of the occupants and providing medical aid as needed. They also communicate with other rescue personnel to ensure that the occupants receive appropriate care and transport to medical facilities.

D. Rescue Team Members:

Rescue team members are responsible for assisting with the rescue operation, including gaining access to the elevator car, securing the occupants, providing additional medical aid, and removing occupants from the elevator car using appropriate techniques and equipment. Team members must also be prepared to respond to any unexpected developments during the rescue operation and communicate effectively with other personnel.

Each of these personnel roles requires specialized training and certification to ensure that responders are prepared to handle the unique challenges and hazards associated with elevator rescue operations. Additionally, effective teamwork and coordination are essential for ensuring that the rescue operation is carried out safely and efficiently.

Training and Certification Requirements

Elevator rescue personnel require specialized training and certification to ensure they can respond quickly and effectively to emergencies. The following are some common training and certification requirements for elevator rescue personnel:

A. Basic Elevator Rescue Training:

This training provides an overview of elevator rescue procedures and protocols, including how to secure an elevator, communicate with occupants, and safely remove individuals from a stalled elevator. Basic training may also cover equipment usage, situational awareness, and hazard identification.

B. Advanced Elevator Rescue Training:

Advanced training builds on the basic training and provides additional knowledge and skills needed to respond to more complex elevator emergencies. This may include training on how to respond to elevator fires, entrapments, and power outages.

C. Ongoing Training and Refresher Courses:

Ongoing training is essential for elevator rescue personnel to maintain their skills and stay up to date on the latest rescue techniques and equipment. Refresher courses may also be required to ensure that personnel is meeting certification requirements and staying current with new developments in elevator technology and rescue procedures.

Importance of Teamwork and Coordination

Effective teamwork and coordination are critical for a successful elevator rescue operation.

Clear communication between all members of the rescue team is essential for ensuring that everyone is aware of the situation and that all necessary steps are taken to address it. This includes communicating with building management, emergency services, and any other involved parties.

Efficient task assignment and prioritization are also crucial for a successful rescue operation. Each member of the rescue team must understand their role and responsibilities, and work together to achieve the common goal of safely rescuing the occupants of the elevator.

Maintaining situational awareness and safety is also a critical aspect of elevator rescue operations. All members of the rescue team must be aware of potential hazards and risks during the operation and take steps to mitigate them.

Post-incident debriefing and analysis is essential for the continuous improvement of elevator rescue operations. By analyzing the response to the emergency, the rescue team can identify areas for improvement and develop strategies for more effective future response.

Conclusion

The history and mechanics of elevator rescue have been thoroughly discussed in this section. We have examined the evolution of elevators and their rescue procedures over time, as well as the various types of elevators and their unique challenges in rescue operations.

Rescue personnel must have a comprehensive understanding of the mechanics of elevators and the history of elevator rescue to be able to respond effectively to elevator emergencies. This knowledge is essential in determining the appropriate rescue strategies, identifying access points, and selecting the right tools and equipment to carry out a successful rescue operation.

By considering the key factors that can impact elevator rescue operations, such as building design, elevator location, number of occupants involved, and the availability of rescue equipment, rescue personnel can prioritize the safety of both occupants and rescuers.

Finally, it is essential to maintain clear communication, coordination, and teamwork during elevator rescue operations. Training and certification programs must be up-to-date, and ongoing training and refresher courses should be provided to keep elevator rescue personnel knowledgeable and efficient.

In conclusion, understanding the history and mechanics of elevator rescue is vital for the success of any rescue operation. By incorporating this knowledge, rescue personnel can respond efficiently and effectively to any elevator emergency, ultimately leading to a safe and successful rescue.

VI. Phase 1: Preliminary Steps

The next three chapters review the 3 phases of the elevator rescue in depth. Firefighters need to have pre-plans for elevators in their area under jurisdiction. In cooperation with building management, they should have conducted rescue drills using those elevators beforehand. However, there will be times when a crew may be deployed to an area that is unfamiliar to them. These are the steps by which we can conduct a rescue.

Elevator rescue operations are time-sensitive and require a well-structured plan to ensure the safety of all occupants and responders involved. Phase 1 is the initial stage of the rescue process, where the first steps are taken to assess the situation and remove occupants from the stalled elevator.

The main objectives of phase 1 are to quickly identify the stalled elevator and its location, assess the situation, and communicate with the occupants. The safety of both the occupants and responders is the top priority during this phase. The responders need to act efficiently and professionally to prevent further harm and damage to the occupants and the elevator system.

A. Identification of the Stalled Elevator and Initial Assessment

When responding to an elevator emergency, the first step is to identify the stalled elevator. The identification process may involve communicating with the building occupants, reviewing building plans, or conducting a visual search of the elevator shaft.

Once the elevator has been identified, the rescue team will need to make an initial assessment of the elevator's condition. This assessment will help the team determine the appropriate course of action for the rescue operation. Some of the factors that may be considered during the initial assessment include:

1. **Elevator Location**: The location of the stalled elevator may impact the ease and safety of the rescue operation. For example, an

elevator that is located on a high floor of a building may require special equipment and techniques to safely rescue the occupants.

2. **Number of Occupants**: The number of occupants trapped in the elevator will impact the rescue operation. The rescue team will need to determine how many people are in the elevator and their condition before determining the appropriate rescue plan.

3. **Elevator Condition**: The condition of the elevator may also impact the rescue operation. For example, if the elevator is stuck between floors or has malfunctioning doors, the rescue team will need to take extra precautions to safely rescue the occupants.

By carefully assessing these and other factors, the rescue team can develop an effective plan for rescuing the occupants of the stalled elevator.

B. Communication with the Occupants

During an elevator rescue operation, clear and effective communication with the occupants is crucial for their safety and well-being. Here are some important procedures to follow for establishing communication with the occupants:

1. **Assess the situation**: Before initiating any communication with the occupants, it is important to assess the situation and understand the conditions of the elevator. This includes determining the number of occupants, their condition, and any potential hazards that may be present.

2. **Introduce yourself**: Once you have assessed the situation and determined that it is safe to do so, introduce yourself to the occupants. Provide your name, your role in the rescue operation, and a brief explanation of what is happening.

3. **Provide reassurance**: Being stuck in an elevator can be a frightening experience for occupants, so it is important to provide reassurance and help them remain calm. Let them know that they are not alone, and that help is on the way.

4. **Obtain information**: Ask the occupants if they are injured or if they have any medical conditions that need immediate attention. This information will help you determine the appropriate course of action and provide medical aid, if necessary.

5. **Explain the rescue procedure:** Provide the occupants with a brief overview of the rescue procedure and what they can expect during the rescue operation. Let them know that they may feel some movement as the elevator is being stabilized, but that it is normal and necessary for their safety.

Effective communication with the occupants can help to minimize their anxiety and stress during the rescue operation and ensure their safety throughout the process.

C. Removal of Occupants from the Stalled Elevator

When rescuing occupants from a stalled elevator, the safety of the occupants should be the top priority. The following are some techniques for safely and efficiently removing occupants from the elevator:

1. **Use a ladder**: If the elevator is not stuck between floors, the occupants may be able to climb out using a ladder. However, this method is only recommended for able-bodied individuals.

2. **Use a pry bar**: If the elevator is stuck between floors, the rescuers can use a pry bar to open the doors and help occupants climb out. However, this should only be done by experienced professionals, as it can be dangerous if not done properly.

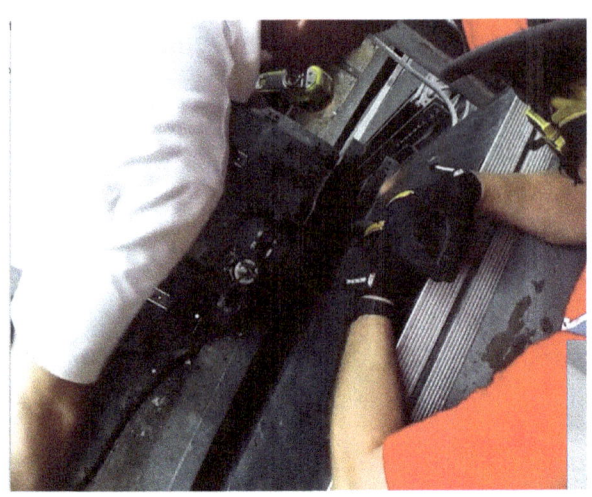

Figure 3: Opening the elevator car door using the pry bar. Interlocks and door mechanisms vary per manufacturer. This is why it is important to pre-plan and conduct drills even before an emergency.

3. **Use an elevator key**: Some elevators have a key that allows the doors to be opened manually. If the elevator has this feature, the rescuers can use the key to open the doors and help occupants climb out.

4. **Use a hoist**: In some cases, a hoist can be used to lift the occupants out of the elevator. This method requires special equipment and should only be done by experienced professionals.

It is important to follow established rescue procedures and protocols when removing occupants from a stalled elevator. Rescuers should communicate clearly and effectively with the occupants throughout the rescue operation, providing clear instructions and reassurance to help keep them calm.

If an occupant has a medical condition or injury, special care should be taken when removing them from the elevator. Rescuers should work with medical personnel to assess the occupant's condition and determine the best method for removing them from the elevator safely and efficiently.

D. Recycling the Power on the Main Disconnect to the Elevator Car

One of the first steps in attempting to free the trapped occupants of a stalled elevator is to attempt to recycle the power on the main disconnect to the elevator car. The main disconnect is an electrical switch that cuts off power to the elevator and recycling it can sometimes reset the elevator's controls and allow it to resume normal operation.

Before recycling the power, it is important to verify that it is safe to do so. Elevator technicians or other trained personnel should take this step, as there are significant risks associated with handling electrical equipment. This includes checking that the elevator car is properly aligned with a landing and that no one is in contact with the elevator or its components.

If it is determined to be safe, the power can be recycled by opening the main disconnect switch, waiting for a few seconds, and then closing the switch again. The elevator car should then be tested to see if it is operational.

It is important to note that recycling the power is not always successful and should not be attempted in all situations. Elevator technicians and other trained personnel should be consulted to determine the appropriate course of action. Additionally, all steps should be taken to ensure the safety of everyone involved in the rescue operation.

E. Conclusion

In this section, we covered phase 1 of elevator rescue operations, which involves identifying the stalled elevator and taking initial steps to ensure the safety of the occupants and responders. We discussed the importance of clear and effective communication with occupants, as well as techniques for safely and efficiently removing them from the elevator.

We also went over the procedures for recycling the power on the main disconnect to the elevator car and the importance of following established safety protocols during this process.

It is essential to follow established procedures to ensure the safety of occupants and responders during elevator rescue operations. In the next section, we will discuss the next phase of the operation, which involves preparing for the actual rescue.

VII. Phase 2: Lock-Out/Tag-Out

Elevator rescue operations involve a systematic approach to ensure the safety of both occupants and responders. Phase 2 of the operation, which involves lock-out/tag-out, is a crucial step toward achieving this goal. This section will provide an overview of phase 2 and its importance in elevator rescue operations, as well as the key objectives of this phase.

A. Definition of Lock-Out/Tag-Out

Lock-out/tag-out is a crucial safety procedure that is used to protect individuals who work on or around machinery, including elevators. It involves the isolation of the equipment's energy source and the placement of a lock or tags on the isolating device to prevent the machine from being energized while maintenance or repair work is being performed. The lock or tag is a physical barrier that is used to signal that the equipment is not safe to use.

The main purpose of lock-out/tag-out is to prevent injuries and accidents that can occur when machinery is accidentally or unexpectedly energized while maintenance or repair work is being performed. This procedure is critical to the safety of both the workers and anyone else who may meet the machinery. It ensures that the machinery cannot be started or used until the lock or tag is removed by an authorized individual.

B. Procedures for Locking Out/Tagging Out the Elevator Machinery

One of the primary objectives of phase 2 of the elevator rescue operation is to ensure that the elevator machinery is safely locked out or tagged out before any work is performed on the elevator car. This is to prevent the accidental activation of the elevator machinery and to safeguard the rescue personnel and the occupants.

1. Identification of the power sources and how to shut them off:

The first step in locking out/tagging out the elevator machinery is to identify the power sources that need to be shut off. Typically, there are two main sources of power to an elevator: electrical power and hydraulic power. It is essential to identify the power sources that are supplying power to the elevator and how to shut them off safely.

2. Procedures for locking out/tagging out the machinery:

Once the power sources have been identified, the next step is to lock out or tag out the machinery. Lock-out/tag-out involves placing a lock or tag on the power source to prevent anyone from inadvertently activating the machinery. The lock or tag must be placed in a position where it is visible and accessible to all rescue personnel.

3. Importance of following established safety protocols:

It is crucial to follow established safety protocols when locking out/tagging out the elevator machinery. This includes using the appropriate lock or tag, following the correct procedures for shutting off the power sources, and ensuring that all personnel involved in the rescue operation are aware of the lock-out/tag-out procedures.

Failure to follow established safety protocols can result in serious injuries or fatalities. Therefore, it is essential to follow the lock-out/tag-out procedures carefully and to ensure that all personnel involved in the rescue operation understand the importance of following established safety protocols.

By following the lock-out/tag-out procedures, rescue personnel can ensure that the elevator machinery is safely locked out or tagged out, and the rescue operation can proceed safely.

C. Identification of the Power Sources

In Phase 2 of elevator rescue operations, it is essential to identify and shut off all power sources to the elevator machinery. This process is crucial in ensuring that the machinery is not accidentally restarted, leading to injuries or accidents during the rescue operation. The following are the different power sources in the elevator machinery that need to be identified:

1. **Electrical power source** - this is the primary power source that runs the elevator motor and controls. It is usually located in the main electrical panel of the building.

2. **Backup power source** - this power source is used in case of a power outage or failure of the primary power source. It is typically located in a separate panel and can be a battery or a generator.

3. **Hydraulic power source** - this power source is used in hydraulic elevators and is responsible for lifting the elevator car using hydraulic fluid. It is usually located in a separate room or machinery space.

Once all power sources have been identified, the next step is to shut them off using established procedures. This process involves cutting off the electrical power supply to the machinery and locking out the machinery using lockout devices or tags.

It is essential to follow established safety protocols when shutting off power sources and locking out the machinery to prevent accidental restart. Proper communication and coordination among the rescue team members are critical during this phase of the operation to ensure that all power sources have been identified and shut off.

D. Removal of the Control Key or Lockout Device

When performing lock-out/tag-out procedures, it is crucial to ensure that the control key or lockout device is removed. The following are procedures for safely removing the control key or lockout device:

1. Identify the control key or lockout device that needs to be removed.

2. Check that all the power sources have been shut off and locked out/tagged out.

3. Verify that the lockout device is secure and cannot be removed accidentally.

4. Remove the control key or lockout device.

5. Store the control key or lockout device in a secure location away from unauthorized personnel.

6. Document the removal of the control key or lockout device.

It is important to note that the lockout device must remain in place until the work is completed and all personnel and equipment are safely clear of the elevator machinery.

E. Testing the Elevator Controls

Once the elevator machinery has been locked out/tagged out, it is essential to test the controls to ensure that the machinery is properly de-energized and cannot be restarted. Here are the procedures for testing the elevator controls:

1. Verify that the control key or lockout device has been removed.

2. Test the elevator controls by attempting to call the elevator from different floors. If the elevator does not respond, it means that the controls have been successfully locked out/tagged out.

3. If the elevator responds to the call, it is important to repeat the lock-out/tag-out procedures and verify that all power sources have been disconnected.

4. Conduct a visual inspection of the elevator machinery to ensure that all hazardous energy sources have been identified and controlled.

5. Once all these procedures have been completed, the elevator machinery is considered safe to work on.

It is important to note that testing the elevator controls should only be done by qualified elevator rescue personnel who have been trained in lock-out/tag-out procedures. Testing the controls can be dangerous if not done properly, so it is crucial to always follow established safety protocols.

E. Conclusion

In this section, we discussed the second phase of elevator rescue operations, which is lock-out/tag-out. We explained the importance of this phase in preventing injuries and accidents during the rescue operation. We also discussed the procedures for locking out/tagging out the elevator machinery, identifying the power sources, removing the control key or lockout device, and testing the elevator controls.

It is crucial to follow established safety protocols during this phase to ensure the safety of all occupants and responders involved in the rescue operation. Failure to do so could result in injuries or even fatalities.

In the next section, we will discuss the final phase of elevator rescue operations, which involves accessing the elevator machinery and performing the necessary repairs to restore the elevator to working condition.

VIII. Phase 3: Car Movement and Exit

Phase 3 of elevator rescue operations involves the safe movement and exit of the stalled elevator car. This phase is critical in ensuring the safety of both the occupants and responders involved in the rescue operation. The key objectives of phase 3 are to safely move the stalled elevator car, prepare the hoistway for movement, provide safe exit procedures for occupants, and recover all equipment used during the rescue operation. In this section, we will discuss the procedures involved in phase 3 of elevator rescue operations.

A. Procedures for Movement of the Stalled Elevator Car

Phase 3 of elevator rescue operations involve the movement and eventual exit of the stalled elevator car. The procedures for moving the car must be carried out with the utmost caution and care to ensure the safety of both the occupants and the responders involved. Some of the key procedures for the movement of the stalled elevator car include:

1. Explanation of the procedures for moving the stalled elevator car:

Rescuers must first assess the situation to determine the best course of action for moving the stalled car. The type of elevator, the location of the car, and the condition of the occupants are all factors that must be considered.

2. Importance of ensuring safety during the movement process:

The safety of the occupants and the rescuers must be the top priority during the movement process. Rescuers must take precautions to prevent the car from falling, tipping, or jerking during the movement.

3. Techniques for controlling the movement of the elevator car:

Several techniques can be used to control the movement of the elevator car, including manual operation of the controls, the use of a secondary power source, and the use of an external hoist or winch. The most appropriate technique will depend on the specific circumstances of the rescue operation.

B. Preparation of the Elevator Hoistway

Procedures for preparing the elevator hoistway for the movement of the car.

Before moving the stalled elevator car, it is important to ensure that the elevator hoistway is prepared and safe for movement. Here are some procedures for preparing the hoistway:

- Verify that the hoistway is clear of any obstructions or debris that may impede the movement of the elevator car.

- Inspect the hoistway for any signs of damage, such as broken or loose rails, damaged guide shoes, or missing bolts or clips. Repair any damages before proceeding with the movement of the elevator car.

- Check the elevator shaft for proper ventilation and lighting. Adequate ventilation and lighting are necessary to ensure the safety of the occupants and responders during the rescue operation.

- Make sure that the elevator pit is free of water or other hazardous materials that may cause injury to the occupants or damage to the elevator machinery.

Importance of verifying that the hoistway is safe for movement.

Verifying that the hoistway is safe for movement is critical in ensuring the safety of the occupants and responders during the elevator rescue operation. Failure to properly inspect and prepare the hoistway may result in accidents, injuries, or damage to the elevator machinery. Therefore, it is important to follow the established procedures for preparing the hoistway before moving the stalled elevator car.

C. Methods for Accessing the Elevator Hoistway

Methods for accessing the elevator hoistway can vary depending on the specific situation, but some general techniques can be used. One common method is to access the hoistway through the elevator doors on the nearest landing to the stalled car. In some cases, responders may need to remove a portion of the hoistway wall or ceiling to gain access.

Responders must have the proper safety equipment and training before attempting to access the hoistway. This may include personal protective equipment, such as gloves and hard hats, as well as specialized tools and training in fall protection and confined space entry. Proper communication and coordination between responders are also crucial to ensure the safety of all involved.

D. Exit Procedures for the Occupants

Procedures for safely and efficiently removing occupants from the elevator car.

Once the elevator car has been moved to a safe location and the hoistway has been prepared for an exit, the next step is to safely remove the occupants from the car. The following procedures should be followed:

- **Determine the condition of the occupants**: Before proceeding with the exit process, it is important to assess the condition of the occupants to ensure that they are stable and able to exit safely.

- **Communicate with the occupants**: The occupants should be informed of the exit process and any safety procedures that they need to follow.

- **Prepare the exit area**: The area where the occupants will exit the elevator should be cleared and made safe.

- **Assist the occupants**: If necessary, responders should aid the occupants as they exit the elevator. This may include helping them to walk, providing a wheelchair or stretcher, or supporting them if they have sustained injuries.

Importance of prioritizing occupant safety during the exit process

During the exit process, the safety of the occupants should be the top priority. Responders should take the time to ensure that the occupants are safe and comfortable before proceeding with the exit process.

Techniques for managing occupants with medical conditions or injuries.

If an occupant has a medical condition or injury, responders should take special care to ensure their safety during the exit process. This may include providing first aid or medical assistance, as well as modifying the exit process to ensure that the occupant is not further injured. For example, if an occupant has a leg injury, they may need to be assisted out of the elevator in a specific manner to avoid aggravating the injury.

Once all occupants have been safely removed from the elevator, they should be assessed for any injuries or medical issues. If necessary, they should be transported to a medical facility for further evaluation and treatment.

E. Final Assessment and Recovery of Equipment

Procedures for assessing the elevator machinery and hoistway.

After the occupants have been safely removed from the elevator, responders should conduct a final assessment of the elevator machinery and hoistway. This assessment should include:

- Checking the elevator controls and safety systems to ensure that they are functioning properly.
- Inspect the elevator hoistway for any damage or safety hazards.
- Ensuring that all lock-out/tag-out devices have been removed and that the elevator is no longer locked out/tagged out.

Recovery of equipment

Once the assessment is complete, all equipment used during the elevator rescue operation should be recovered and returned to their proper storage locations. Any equipment that requires cleaning or maintenance should be properly cleaned and maintained before being returned to service.

F. Conclusion

In phase 3 of the elevator rescue operation, the stalled elevator car is moved to a safe location, the hoistway is prepared for an exit, the occupants are safely removed from the elevator, and the elevator machinery and hoistway are assessed.

Following established procedures during an elevator rescue operation is essential to ensure the safety of all involved. Responders should be properly trained and equipped to carry out the procedures outlined in each phase of the rescue operation.

IX. Firefighter Training Guidelines

Elevator rescue operations can be complex and dangerous situations that require firefighters to have a specific set of skills and knowledge. Responders must be prepared to handle various types of elevators, including hydraulic, traction, and machine-room-less elevators, as well as various modes of operation, such as automatic, semi-automatic, and manual.

Firefighters must also be trained on how to manage occupants and communicate effectively with them during the rescue operation. Proper training on lock-out/tag-out procedures is crucial to prevent accidental movement of the elevator car, which can result in serious injury or death to both occupants and responders. Additionally, proper techniques for accessing the elevator hoistway, controlling elevator movement, and safely removing occupants are essential to minimize the risk of injury or death to all parties involved.

Firefighters also need to be trained to work efficiently as a team and communicate effectively with other first responders, including police and medical personnel. Teamwork is essential for successful elevator rescue operations, and training should focus on developing the skills and knowledge necessary to work collaboratively in high-stress situations.

Overall, firefighter training guidelines are critical to ensuring that responders are equipped with the necessary skills and knowledge to handle elevator rescue operations safely and efficiently. Without proper training, responders risk their safety as well as that of the occupants they are trying to rescue.

The firefighter training guidelines are a set of standards and protocols that are designed to provide comprehensive training to firefighters, so they can respond to elevator emergencies effectively. These guidelines cover a range of topics, including the identification of different types of elevators, the function and operation of elevator machinery, and the protocols for handling elevator malfunctions and emergencies.

By adhering to these guidelines, firefighters can ensure that they have the necessary skills and knowledge to manage and control the situation safely and efficiently. The guidelines also ensure that all firefighters are trained to a consistent standard, regardless of their level of experience, and that they are prepared to handle any situation that arises.

The guidelines not only provide the necessary training to deal with elevator rescue operations, but they also ensure that firefighters understand the importance of communication, teamwork, and situational awareness. The guidelines are updated regularly to reflect new developments in elevator technology and rescue techniques, ensuring that firefighters remain up to date with the latest methods and practices.

Ultimately, the firefighter training guidelines are critical to ensuring the safety of both the occupants of the elevator and the responding firefighters. Proper training ensures that firefighters are equipped to handle any situation that arises, which can help to prevent injuries, fatalities, and property damage.

In this section, we will discuss the importance of firefighter training guidelines and the key objectives that these guidelines aim to achieve.

A. Recruit Firefighter Training Guidelines

Recruit firefighter training guidelines are a set of standards that specify the minimum training requirements for newly hired firefighters in elevator rescue operations. These guidelines aim to ensure that recruits receive a thorough education in elevator rescue operations, as well as other firefighting disciplines. Providing comprehensive training to recruits is crucial because it lays the foundation for their future development as firefighters.

The recruit firefighter training guidelines provide an essential foundation of knowledge for new firefighters who may not have previous experience with elevator rescue operations. Understanding the different types of elevators and their associated hazards is critical for firefighters to assess the situation quickly and develop an appropriate response. Proper use of rescue equipment, such as door and car opening tools, is essential to ensure a safe and efficient rescue.

In addition to technical skills, recruits may also receive instruction on teamwork and communication with other first responders. Elevator rescue operations often involve coordination between firefighters, building staff, and emergency medical services. Therefore, recruits may receive training on how to communicate effectively with different parties involved in the rescue operation.

Recruit training may also include procedures for incident reporting and documentation. Documenting the rescue operation is crucial for accountability and future reference. Therefore, recruits may learn how to document incidents accurately and appropriately.

Overall, providing comprehensive training to recruits is essential to ensure that they have the necessary skills and knowledge to respond to elevator emergencies safely and efficiently.

It is essential to follow these guidelines to ensure that all recruits have the necessary skills and knowledge to perform their job safely and effectively. Proper training can minimize the risk of injury to firefighters and occupants and increase the likelihood of a successful rescue operation.

B. In-House Firefighter Training Guidelines

In-house firefighter training guidelines are designed to provide ongoing training and education to firefighters to maintain and improve their skills and knowledge in elevator rescue operations. The importance of ongoing training cannot be overstated as it helps ensure that firefighters remain up to date with the latest techniques, technologies, and best practices.

In-house firefighter training guidelines provide more advanced and specialized training to experienced firefighters who have already completed their initial training. The purpose of in-house training is to ensure that firefighters remain up to date with the latest techniques, technologies, and procedures relevant to elevator rescue operations.

Advanced elevator rescue techniques covered in in-house training guidelines may include techniques for rescuing occupants from high-rise elevators or elevators that are in hazardous or hard-to-reach areas. Situational awareness training may help firefighters develop the skills necessary to assess a situation and make informed decisions about the best course of action.

Risk assessment training can help firefighters understand how to identify and assess hazards associated with elevator rescue operations and develop strategies for mitigating these risks. Incident command procedures training can help firefighters learn how to manage an incident effectively and efficiently, ensuring that all personnel are working together to achieve a successful outcome.

In addition, in-house training guidelines may cover topics related to the proper use and maintenance of elevator rescue equipment, such as rescue slings, emergency power systems, and ventilation equipment. Firefighters may also receive training on how to communicate and coordinate with other first responders, such as police officers and paramedics.

Training on how to handle specific types of elevator emergencies, such as entrapments caused by power outages, fires, or mechanical failures, can also be included in in-house training guidelines. Emergency medical procedures training can help firefighters learn how to provide basic first aid and emergency medical treatment to occupants who may have sustained injuries during an elevator emergency.

Overall, in-house firefighter training guidelines should be designed to provide firefighters with the knowledge, skills, and confidence they need to perform elevator rescue operations safely, efficiently, and effectively in a range of emergencies.

C. Conclusion

In conclusion, firefighter training guidelines are crucial to ensuring that firefighters are well-equipped to handle elevator emergencies safely and effectively. The guidelines provide a comprehensive approach to elevator rescue operations, covering a wide range of topics from basic techniques to advanced rescue procedures.

It is important to emphasize that ongoing training and education are essential to maintaining proficiency in these skills, and in-house training guidelines can play a significant role in this regard. Consistent training can also help to identify areas for improvement, such as equipment maintenance and communication protocols, and ensure that the latest procedures and techniques are being utilized.

In summary, by following established training protocols, firefighters can help ensure the safety of occupants and responders during elevator emergencies.

X. Debriefing Procedures for Firefighters

Debriefing procedures provide firefighters with the opportunity to reflect on the incident and evaluate their performance, identifying what went well and what could have been done differently. This process allows for an open and honest discussion among responders to discuss any issues or concerns that arose during the rescue operation. By analyzing the incident and discussing it in a group setting, firefighters can learn from their mistakes and build on their successes, improving their skills and knowledge for future incidents.

Moreover, debriefing procedures are vital for promoting mental wellness among firefighters. Elevator rescue operations can be highly stressful and traumatic events, and debriefing sessions offer responders the chance to process their emotions and share their experiences in a supportive environment. By providing a platform for responders to discuss any emotional or psychological issues they may be experiencing, debriefing sessions can help prevent post-traumatic stress disorder (PTSD) and other mental health conditions.

Debriefing procedures can provide a platform for firefighters to discuss and analyze the incident in a structured and supportive environment. This can help them to process their emotions, reactions, and actions during the rescue operation, and identify any issues or challenges they faced. Through open and honest discussions, firefighters can share their perspectives and experiences, as well as offer feedback to each other on what went well and what could be improved.

Debriefing procedures can also promote a culture of continuous learning and improvement. By encouraging firefighters to reflect on their actions and decisions, debriefing sessions can help to identify gaps in their knowledge, skills, or equipment, and highlight areas where additional training or resources may be needed. This information can be used to update and enhance training protocols and emergency response plans, thereby improving overall safety and effectiveness in future incidents.

Finally, debriefing procedures can be an important component of promoting mental health and well-being among firefighters. Elevator rescue operations can be stressful and traumatic experiences, and debriefing sessions can provide a supportive and non-judgmental space for firefighters to process and cope with any emotional or psychological impacts of the incident. By prioritizing the mental health and well-being of firefighters, debriefing procedures can help to reduce the risk of long-term mental health issues, such as post-traumatic stress disorder (PTSD), and promote overall resilience and job satisfaction.

A. Importance of Debriefing Procedures

Debriefing procedures are important for firefighters because they allow them to reflect on the incident and identify ways to improve their response to similar situations in the future. Through a structured debriefing process, firefighters can evaluate their actions and decision-making during the incident, assess the effectiveness of their response, and identify areas where improvements can be made.

In addition to promoting learning and improvement, debriefing procedures can also have psychological benefits for firefighters. Elevator rescue operations can be stressful and traumatic events, and debriefing sessions can help reduce the impact of these experiences on firefighters' mental health and well-being. By providing a safe and supportive environment to discuss the incident and its aftermath, debriefing procedures can help firefighters process their emotions and develop coping strategies for future incidents.

Moreover, debriefing procedures can foster a sense of camaraderie and teamwork among firefighters. By sharing their experiences and insights, firefighters can build stronger relationships and enhance their ability to work together effectively in future incidents.

B. Guidelines for Effective Debriefing Sessions

Effective debriefing sessions are crucial for promoting learning and continuous improvement among firefighters. Here are some guidelines for conducting effective debriefing sessions:

1. Create a safe and supportive environment for debriefing:

Debriefing sessions should be conducted in a private and confidential setting where firefighters can share their experiences and feelings without fear of judgment or reprisal. It is important to create a safe and supportive environment where firefighters feel comfortable sharing their thoughts and ideas.

2. Facilitate open and honest communication among firefighters:

The facilitator should encourage all participants to share their perspectives and experiences related to the incident. Active listening skills should be employed to understand the different viewpoints and ideas shared by the participants. The facilitator should also encourage questions, clarifications, and suggestions from the participants.

3. Identify areas for improvement and implement changes:

Debriefing sessions should focus on identifying areas for improvement and developing action plans to address them. The facilitator should guide the discussion toward identifying the root causes of the issues and developing actionable recommendations for improvement. The action plans should be clear, concise, and have specific timelines for implementation. The facilitator should follow up with the participants to ensure that the action plans are implemented, and progress is tracked.

Overall, effective debriefing sessions should be conducted in a respectful, supportive, and constructive manner, with a focus on learning and continuous improvement. By implementing the recommendations identified during debriefing sessions, firefighters can enhance their skills, improve their performance, and provide better service to the public.

C. Conclusion

In conclusion, debriefing procedures play a crucial role in ensuring that firefighters can reflect on their performance during elevator rescue operations, identify areas for improvement, and share their experiences with other responders. By creating a safe and supportive environment for open and honest communication, debriefing can help reduce stress and promote overall mental health among firefighters.

To facilitate effective debriefing sessions, it is important to establish guidelines for creating a safe environment, encouraging open communication, and identifying areas for improvement. By incorporating debriefing procedures into regular training and response protocols, firefighters can continue to improve their skills and knowledge, ultimately leading to safer and more efficient elevator rescue operations.

XI. Codes Related to Elevator Rescue

Elevator rescue operations require a high level of technical knowledge and specialized equipment to be performed safely and effectively. The potential hazards associated with these operations can include entrapment, electrical shock, mechanical failure, and fire. To minimize these risks, codes related to elevator rescue have been established to guide best practices and safety measures.

These codes provide minimum safety requirements for the construction, operation, inspection, and maintenance of elevators and related equipment. They also outline procedures for responding to elevator emergencies, including protocols for evacuating occupants and coordinating with other emergency responders. By following these codes, firefighters and other first responders can ensure that they are performing elevator rescue operations safely and effectively.

Codes related to elevator rescue operations are created to set specific guidelines, standards, and minimum safety requirements for firefighters and other first responders to follow when responding to elevator emergencies. These codes provide a framework for the safe and effective operation of elevators during rescue operations, and they ensure that all first responders are operating with a shared understanding of best practices.

The codes outline specific procedures and techniques for safely removing occupants from stalled elevators, accessing the elevator hoistway, and utilizing specialized rescue equipment. They also guide communication and coordination with other first responders and the proper documentation of elevator rescue operations.

Overall, codes related to elevator rescue operations aim to minimize risks and ensure the safety of occupants and responders during elevator emergencies. They are essential tools for firefighters and other first responders to follow and reference during elevator rescue operations.

A. NFPA Codes Related to Elevator Rescue

1. Explanation of the National Fire Protection Association (NFPA)

The National Fire Protection Association (NFPA) is a global nonprofit organization that develops and publishes codes and standards related to fire safety and other hazards. The organization is widely recognized as a leading authority on fire prevention and safety, and its codes and standards are widely adopted and enforced by governments, industries, and organizations around the world.

The NFPA is committed to reducing the risk of fire and other hazards by providing information, education, and codes and standards development. Its codes and standards are developed through a consensus-based process, which involves input from stakeholders in the industry, government, and other interested parties.

The NFPA codes and standards cover a wide range of topics, including fire prevention, life safety, electrical safety, and hazardous materials. They are designed to establish minimum safety requirements, procedures, and best practices to reduce the risk of harm to people, property, and the environment.

The NFPA's codes related to elevator rescue are critical in ensuring that firefighters and other first responders are adequately prepared to handle emergencies involving elevators. These codes guide the proper procedures, techniques, and equipment necessary to safely evacuate occupants from elevators during an emergency. They also cover other important topics, such as communication and coordination with other first responders, risk assessment, and incident command procedures.

2. Overview of the NFPA codes related to elevator rescue

The NFPA has developed several codes related to elevator rescue operations that are designed to ensure the safety of both occupants and responders during emergencies. These codes guide specific safety requirements, procedures, and best practices that firefighters and other first responders must follow when responding to elevator emergencies.

One of the most important codes related to elevator rescue is the NFPA 70: National Electrical Code. This code outlines the electrical safety requirements for elevators, including wiring and grounding, to prevent electrical hazards and ensure safe operation.

Another key code is the NFPA 101: Life Safety Code, which sets forth requirements for life safety in buildings and structures. This code includes specific provisions for elevator rescue, such as the installation of emergency communication systems and the availability of alternate means of escape from buildings.

The NFPA 1221:
The standard for the Installation, Maintenance, and Use of Emergency Services Communications Systems is another important code related to elevator rescue. This code establishes requirements for emergency communication systems that enable firefighters and other responders to communicate during emergencies, including elevator rescue operations.

In addition, the NFPA 130:
Standard for Fixed Guideway Transit and Passenger Rail Systems provides requirements for the design and operation of fixed guideway transit systems, which include elevators and escalators.

Finally, the NFPA 502:
Standard for Road Tunnels, Bridges, and Other Limited Access Highways provides requirements for the design and operation of tunnels, bridges, and other limited access highways, which may include elevators and other vertical transportation systems.

By adhering to these codes, firefighters and other first responders can ensure that they are following established safety protocols and best practices when responding to elevator emergencies, thus reducing the risk of injury or death to occupants and responders.

3. Explanation of the specific requirements outlined in the NFPA codes related to elevator rescue

The NFPA codes related to elevator rescue operations have specific requirements to ensure the safety of occupants and responders during elevator emergencies. One key aspect of the codes is the electrical safety and control systems for elevators, which aim to prevent electrical hazards and ensure the proper functioning of the elevator during rescue operations.

The codes also cover emergency communication systems and procedures, including requirements for communication between the occupants and the rescue team, as well as communication among the rescue team members themselves. This is crucial for effective coordination and quick response during elevator emergencies.

The codes also guide the training and qualifications required for elevator rescue operations. This includes the knowledge and skills necessary for assessing the situation, determining the best course of action, and using the equipment and tools required for the rescue operation.

The codes also specify the equipment and tools needed for elevator rescue operations, including the types of ropes, harnesses, and other rescue gear. These requirements ensure that the rescue team is adequately prepared and equipped to handle any situation that may arise.

Additionally, the codes outline the procedures for safely removing occupants from stalled elevators, including the use of ladders, hoists, and other techniques. The codes also provide evacuation procedures and requirements for high-rise buildings and other structures, which are crucial for the safety of occupants during emergencies.

Finally, the codes emphasize the importance of coordination and communication among first responders during elevator rescue operations. This includes communication with other responding agencies and organizations, as well as clear procedures for handing off the rescue operation to other responders as necessary.

By adhering to these codes, firefighters and other first responders can ensure that they are following established safety protocols and best practices when responding to elevator emergencies, thus reducing the risk of injury or death to occupants and responders.

B. International Guidelines Related to Elevator Rescue

The International Organization for Standardization (ISO) is an independent, non-governmental international organization that develops and publishes standards for various industries, including elevators and elevator rescue operations. ISO has developed several standards related to elevator rescue operations, including ISO 18738:2012, which provides guidelines for the safe and effective rescue of occupants from stalled elevators.

The European Committee for Standardization (CEN) is another international organization that develops and publishes standards for various industries, including elevators and elevator rescue operations. CEN has developed several standards related to elevator rescue operations, including EN 81-28:2003, which provides requirements and recommendations for the safe and efficient rescue of persons trapped in elevators.

The International Association of Elevator Engineers (IAEE) is a professional organization that provides education and training to elevator engineers and technicians. The organization has developed several guidelines related to elevator rescue operations, including best practices for the safe and efficient rescue of occupants from stalled elevators.

These international guidelines aim to provide additional guidance for firefighters and other first responders, especially in countries where NFPA codes are not adopted or enforced. While there may be differences in the specific requirements outlined in these guidelines compared to NFPA codes, the overarching objective of ensuring the safety of occupants and responders during elevator rescue operations is the same.

The ISO has published several standards related to elevator rescue operations, including:

ISO 18738:2012, titled "*Emergency lighting,*" provides guidelines for emergency lighting in buildings, including elevators, to ensure that emergency lighting is available and functioning during power failures or other emergencies.

ISO 22200:2018, titled "*Lifts (elevators), escalators, and moving walks - Risk assessment and reduction methodology,*" guides the risk assessment process for lifts, escalators, and moving walks, and establishes procedures for identifying and assessing risks related to these systems.

ISO 25745-2:2015, titled "*Energy performance of lifts, escalators and moving walks - Part 2: Energy calculation and energy efficiency benchmarks for lifts (elevators),*" provides a method for calculating the energy performance of elevators, and sets energy efficiency benchmarks for elevators.

These ISO standards guide different aspects of elevator rescue operations, including safety, risk assessment, and energy efficiency.

The CEN has also published several standards related to elevator rescue operations, including:

EN 81-28:2003 specifies safety rules for the construction and installation of lifts and requires that remote alarms are fitted on passenger and goods passenger lifts installed in buildings. The remote alarm is intended to inform the rescue services of the lift location and to enable them to take appropriate action in the event of an emergency.

EN 81-72:2015 provides safety rules for the construction and installation of firefighter lifts that are designed to be used exclusively by firefighters. This standard specifies the technical and safety requirements for the installation, testing, and maintenance of firefighter lifts, including the design of the lift car, the installation of control devices, and the use of emergency lighting and communication systems.

EN 81-20:2014 provides safety rules for the construction and installation of passenger and goods passenger lifts, including design, construction, and testing requirements. This standard also includes requirements for emergency lighting, ventilation, and communication systems in the lift car and the lift shaft.

The International Association of Elevator Engineers (IAEE) has also developed guidelines related to elevator rescue operations. These guidelines provide valuable information and best practices for firefighters and other first responders when responding to elevator emergencies.

One of the IAEE guidelines related to elevator rescue is the "IAEE Guidelines for Rescue from Elevators." This guideline provides an overview of the basic procedures and considerations for rescuing occupants from stalled elevators, as well as best practices for communication with occupants and coordination with other first responders.

Additionally, the IAEE has developed a "Recommended Practice for Responding to Elevator Emergencies." This document provides more detailed guidance on elevator emergency response, including procedures for elevator entrapment, rescue, and evacuation. It also covers topics such as elevator control systems, communication systems, and the use of specialized equipment and tools.

These IAEE guidelines serve as valuable resources for firefighters and other first responders, providing guidance and best practices for ensuring the safety of occupants and responders during elevator rescue operations.

These international guidelines provide additional guidance and best practices for elevator rescue operations. They cover topics such as risk assessment, communication, coordination with other responders, and equipment and tools.

The NFPA codes and international guidelines related to elevator rescue operations are developed by different organizations and are based on different regulatory frameworks, which can result in some differences in their specific requirements and recommendations. However, despite these differences, both the NFPA codes and international guidelines share a common goal of ensuring the safety of both occupants and responders during elevator rescue operations.

For example, the NFPA codes focus on the specific requirements related to electrical safety and control systems for elevators, emergency communication systems and procedures, training and qualifications for elevator rescue operations, and equipment and tools for rescue operations. On the other hand, the ISO, CEN, and IAEE guidelines cover a broader range of topics, including risk assessment and reduction, energy efficiency, and recommended practices for responding to elevator emergencies.

Regardless of the differences in focus and scope, both the NFPA codes and international guidelines provide important guidance and best practices for firefighters and other first responders who are responsible for responding to elevator emergencies. By following established codes and guidelines, responders can ensure that they are well-prepared to handle any situation that arises and can help minimize the risks and hazards associated with elevator rescue operations.

C. Conclusion

In conclusion, codes related to elevator rescue operations play a critical role in ensuring the safety of both occupants and first responders. The National Fire Protection Association (NFPA), as well as international organizations such as the International Organization for Standardization (ISO), the European Committee for Standardization (CEN), and the International Association of Elevator Engineers (IAEE), have developed guidelines and standards to provide comprehensive guidance for firefighters and other first responders during elevator rescue operations.

These codes and guidelines cover a wide range of topics, including electrical safety, emergency communication systems, training and qualifications, equipment and tools, evacuation procedures, and coordination with other first responders. Following these established codes is crucial to ensure the safety of occupants and responders during elevator rescue operations.

In summary, elevator rescue operations pose significant risks, and codes related to elevator rescue provide minimum safety requirements, procedures, and best practices to guide responders during these operations. It is essential to adhere to these established codes to ensure the safety of all parties involved.

About the Author

Easy to spot me as I squeeze through for the picture. I'm the 7th in the front row from the left.

Zero Mella is a practicing physician and a fire training instructor in the Philippines. He has been actively involved in various health programs, including Disaster Risk Reduction and Management on Health for the City of Olongapo. In 2016, he was part of a six-man delegation of firefighters enrolled in the Virginia Beach Fire Academy as part of a Sister City Agreement.

After graduating from the academy, Zero was sent back to Olongapo to teach as a Fire Training Instructor. Although he had to divide his time between being a rural physician and a fire instructor, he found writing firefighting-related articles a niche. This book is a collection of his research throughout the years, aimed at providing guidance and best practices for firefighters and other first responders involved in elevator rescue operations.

www.ingramcontent.com/pod-product-compliance
Lightning Source LLC
Chambersburg PA
CBHW070429180526
45158CB00017B/931